Insects & Bugs
of North America

Jaret C. Daniels

Adventure Quick Guides
YOUR WAY TO EASILY IDENTIFY INSECTS & BUGS

Adventure Quick Guides

How many times have you seen a bug and wondered, "What in the world is that?" This Adventure Quick Guide provides an easy and fun way to identify them. It features 136 of the most common and readily observable insects and bugs in North America. Whether you're exploring a wilderness area, a nearby park or vacant field, or even your own backyard, this guide has you covered.

With its great diversity of climates, habitats, and ecosystems, North America supports a tremendous diversity of insects. There are well over 150,000 insect species found within the United States and Canada alone, and many more species of "bugs" if other arthropods, the larger group to which insects, spiders, centipedes, and scorpions all belong, are included. So get outside and start looking. You'll be amazed at what you can find!

Cover and book design by Lora Westberg
Edited by Brett Ortler

Anatomy Illustration by Julie Martinez
Cover image: Cathy Keifer/Shutterstock.com

All images copyrighted.

All photos by Jaret C. Daniels unless otherwise noted.

Photos identified by page in descending positions a, b, c, d, e, f
Ray Bruun: 6e
Photos by Shutterstock.com:
Evgeniy Ayupov: 21e **AZ Outdoor Photography:** 11b **Nancy Bauer:** 31e **Gualberto Becerra:** 9e **Randy Bjorklund:** 7b, 9c, 23d **prasom boonpong:** 27c **Boyce's Images:** 23e **Ziga Camernik:** 26d **chanus:** 5c **J. T. Chapman:** 9d **Katarina Christenson:** 12a, 16d **Jansen Chua:** 28d **Mircea Costina:** 11d **crystaltmc:** 27b **Gerald A. DeBoer:** 5a, 5d **David Desoer:** 28a **Laura Dinraths:** 28c **Dennis W Donohue:** 27e **enterlinedesign:** 15b **Melinda Fawver:** 13d, 19a, 21f, 24a **K Quinn Ferris:** 11a **Charlie Floyd:** 19e **Steven Fowler:** 8c **Tom Franks:** 10a **Adrienne G:** 9b **Dmitri Gomon:** 13a **Aleksandar Grozdanovski:** 28b **Andreas H:** 8a **Elliotte Rusty Harold:** 5e, 6c, 12b, 17e, 25b **IrinaK:** 32b **Maria Jeffs:** 7a **Matt Jeppson:** 19f **Jgade:** 16e **Jumos:** 7d **David Byron Keener:** 10d **Cathy Keifer:** cover, 10c, 14c **Breck P. Kent:** 18c **khlungcenter:** 15d, 25a **Mirek Kijewski:** 32a **Anton Kozyrev:** 23a **Pavel Krasensky:** 26a, 30a **D. Kucharski K. Kucharska:** 14d **Jim Lambert:** 29b **DM Larson:** 4b **Henrik Larsson:** 12d, 13c, 26c, 28e **Nikolajs Lunskijs:** 8e **Macrolife:** 16a **Cosmin Manci:** 17b, 32d **markh:** 16b **VINU MATHEW:** 4a **Ian Maton:** 19c, 19d **NatalieJean:** 21c **natfu:** 29a **Jack Nevitt:** 11c **Matee Nuserm:** 29e **Sari Oneal:** 4e, 6d, 12c, 31d **Jukka Palm:** 18f **Cristina Romero Palma:** 21d, 26b **watchara panyajun:** 27d **Warren Parker:** 18a **Paul Reeves Photography:** 4d, 5b, 7c, 13b, 14b, 23b **Kamphol Phorangabpai:** 15e **PHOTO FUN:** 29c **QueSeraSera:** 26e **RachelKolokoffHopper:** 29d **Ron Rowan Photography:** 18b **Norjipin Said:** 21b **Sergio Schnitzler:** 17d **SIMON SHIM:** 17a, 18e **Steven R Smith:** 8d, 14a **Spok83:** 30b **Stana:** 22a **Alex Staroseltsev:** 15a **Josef Stemesede:** 8b **TTstudio:** 23c **Marco Uliana:** 15c **Frances van der Merwe:** 16c **Marsha Wood:** 31c **Pan Xunbin:** 24b **yhelfman:** 9a **Bildagentur Zoonar GmbH:** 10b, 22b

10 9 8 7 6 5 4 3

Insects & Bugs of North America
Copyright © 2019 by Jaret C. Daniels
Published by Adventure Publications, an imprint of AdventureKEEN
310 Garfield Street South, Cambridge, Minnesota 55008
(800) 678-7006
www.adventurepublications.net
All rights reserved
Printed in China
ISBN 978-1-59193-818-7 (pbk.)

JUST HOW MANY BUGS ARE THERE?

In everyday language, we often refer to insects, spiders, and other creepy-crawly organisms, such as centipedes, as bugs. Technically, they are all arthropods, which means they belong to the phylum Arthropoda. Arthropoda is the largest group in the animal kingdom, and it contains some of the most famous groups of animals on the planet, including insects, spiders, millipedes, and crustaceans. All told, arthropods represent approximately 75 percent of all known animal species on Earth. The vast majority of arthropods are insects. What's more, many animal species have yet to be identified, and the vast majority of them are likely insects or other arthropods. In fact, scientists conservatively estimate that there may be more than 8 million insect species on Earth.

Insect Anatomy

Insects have a hard exoskeleton on the outside of their bodies; this provides both protection and support. Their body is divided into three distinct regions: the head, thorax, and abdomen.

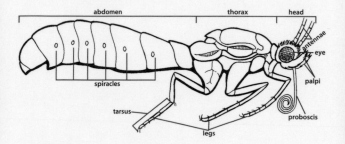

The Head

An insect's head has two prominent compound eyes, two antennae, and mouthparts. The rounded compound eyes are made up of hundreds or thousands of tiny individual eyes, giving insects fairly good vision. The antennae are sensory structures that help with orientation, smell, and taste. The head also bears mouthparts, which vary considerably across insect groups.

The Thorax

The thorax is the middle portion of an insect's body. All insects have three pairs of jointed legs on their thorax, one pair on each segment. Many insects also have one or two pairs of wings, which enable flight and serve a variety of other functions, such as self-defense and mimicry.

The Abdomen

The abdomen contains the reproductive, digestive, and excretory systems along with a series of small lateral holes, called spiracles, that help insects breathe. In female insects, the tip of the abdomen may have an added structure called an ovipositor, which is used to insert or place eggs. In some insects, such as bees and wasps, the ovipositor is modified into a stinger that can be used for self-defense.

WHERE TO FIND INSECTS

Insects and arthropods can be found virtually anywhere, and they are especially common near locations where people live and work. In other words, there are a great many insects and other arthropods around to discover and enjoy. However, due to their small size and often secretive habits, many often go unnoticed—that is, unless you know how and where to look. Observing and studying these wonderful creatures is like opening a treasure chest of natural history. You will quickly discover a hidden world filled with an amazing variety of form and function, including many unique interactions and bizarre behaviors. What is most exciting, though, is that this fascinating world is just outside your front door!

HOW TO USE THIS GUIDE

This guide is organized by where you're likely to encounter specific insects and their relatives. There are six main sections: At Light, In or Near Water, In the Air, On Flowers, On Structures/On the Ground, and on Vegetation. Of course, these categories aren't definitive, but they are a good way to start identifying the organisms you find.

Within each section, the insects and other arthropods are organized by the common name for their scientific order (e.g. beetles, butterflies, and moths, etc.), followed the insect's common range and how large average specimens are. (An order is a scientific rank of similar organisms.)

As an added value, we've also included a handful of notable caterpillars that bug-hunters may encounter outdoors.

A SAFETY NOTE

Be cautious. Some of the organisms identified in this guide can bite, sting, pinch, or otherwise cause irritation, especially if handled or molested. This may result in temporary pain, redness, itching, or minor swelling. If you are allergic to bites or stings, do not closely approach or handle insects or other arthropods.

Western Pygmy-Blue
(Brephidium exilis)

Range: Much of the western U.S., except the far-northern states.

Wingspan 0.5–0.75 of an inch; brown above with gray-blue at wing bases; hindwing below brown with a gray-white base and small black spots along the margin

Melissa Blue
(Plebejus melissa)

Range: The western U.S. and parts of Midwest and Northeast

Wingspan 0.75–1.2 inches; male bright blue above; female brown with orange border above; hindwing below whitish-gray with scattered black spots and prominent orange-spotted band

Dainty Sulphur
(Nathalis iole)

Range: The southern two-thirds of the U.S.

Wingspan 0.75–1.25 inches; wings above yellow with black forewing tip and bar along lower edge; hindwing below yellow with greenish scaling; coloration varies seasonally

Tawny-edged Skipper
(Polites themistocles)

Range: Most of the U.S.

Wingspan 0.8–1.2 inches; wings below brown with distinct contrasting orange scaling along the forewing margin

Common Sootywing
(Pholisora catullus)

Range: Throughout the U.S.

Wingspan 0.9–1.25 inches; shiny dark brown to black; forewing with variable number of white spots near tip; head with white spots

Butterflies and Moths (Lepidoptera)

Peck's Skipper
(Polites peckius)

Range: Most of the U.S.

Wingspan 1.0–1.25 inches; wings below dark brown with distinctive irregular central golden spots

Delaware Skipper
(Anatrytone logan)

Range: The eastern U.S.

Wingspan 1.0–1.4 inches; wings above orange with dark borders and veins; unmarked golden orange below

Little Yellow
(Pyrisitia lisa)

Range: Most of the eastern U.S.

Wingspan 1.0–1.6 inches; wings above bright yellow to whitish with black forewing tip and borders; wings below yellow to whitish with darker markings; hindwing seasonally variable

Common Checkered-Skipper/White Checkered-Skipper
(Pyrgus communis/Pyrgus albescens)

Range: Throughout the U.S.

Wingspan up to 1.25 inches; wings black with numerous white patches giving an overall checkered pattern

Spring Azure
(Celastrina ladon)

Wingspan up to 1.25 inches; pale blue above with dark markings; dusky gray below with varying degrees of dark marks and scaling

Range: Most of the U.S.

American Copper
(Lycaena phlaeas)

Range: Northern two-thirds of eastern U.S.

Wingspan up to 1.4 inches; bright reddish-orange forewings with scattered black spots and hindwings with scalloped orange border

Gray Hairstreak
(Strymon melinus)

Wingspan up to 1.5 Inches; slate gray above with orange-caped black hindwing spot; light gray below with black-and-white dashed lines and orange hindwing spot near tail

Range: Throughout the U.S.

Plume Moth
(Pterophoridae family)

Range: Throughout the U.S.

Wingspan up to 1.5 inches; slender, elongated body; dull-colored narrow wings that resemble a T shape when at rest

Juvenal's Duskywing
(Erynnis juvenalis)

Range: The eastern U.S.

Wingspan 1.5–1.9 inches; dark brown to blackish; forewing pattered with gray and brown and small clear spots near the tip; female more heavily patterned than male

Pacific Orangetip
(Anthocharis sara)

Wingspan up to 1.5 inches; wings white to yellowish-white with orange patch with black markings at forewing tip

Range: West Coast of the U.S.

Butterflies and Moths (Lepidoptera)

Great Purple Hairstreak
(Atlides halesus)

Wingspan up to 1.7 inches; metallic blue above; dull black below with red markings and long hindwing tails; bright orange abdomen

Range: The southern half of the U.S.

American Snout
(Libytheana carinenta)

Range: The eastern and southwestern U.S.

Wingspan up to 1.9 inches; brown above with orange patches and white forewing patches; variable brown below; forewings squared off; elongated labial palps (scaly lobes on both sides of the proboscis) create a snout-like appearance

Clouded Sulphur
(Colias philodice)

Wingspan 1.9–2.5 inches; wings above clear lemon yellow with black borders and dark forewing spot; hindwing below greenish yellow with silvery central spots; females also have a white form

Range: Most of the U.S., except much of California

Cabbage White
(Pieris rapae)

Wingspan up to 2.0 inches; wings white with black forewing tips and black spots

Range: Throughout the U.S.

Checkered White
(Pontia protodice)

Wingspan up to 2.0 inches; wings white with charcoal spots and markings

Range: Throughout the U.S.

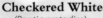

Butterflies and Moths (Lepidoptera)

ON FLOWERS

Range: Throughout the U.S.

Hummingbird Moth
(Genus *Hemaris*)

Wingspan up to 2.2 inches; stout, hairy body with golden thorax and dark abdomen; partially transparent wings

Common Buckeye
(*Junonia coenia*)

Wingspan up to 2.25 inches; brown above with large eyespots; white forewing patch and orange bars; variable brown to reddish below

Range: Throughout most of the U.S.

Range: Throughout the U.S.

Silver-spotted Skipper
(*Epargyreus clarus*)

Wingspan up to 2.4 inches; stout body; wings brown with a distinct white patch on hindwing below

Range: Throughout the U.S.

Orange Sulphur
(*Colias eurytheme*)

Wingspan up to 2.4 inches; bright yellow-orange above with black borders; wings below yellow with one or two central red-rimmed silvery spots

Range: Throughout the U.S.

Red Admiral
(*Vanessa atalanta*)

Wingspan up to 2.5 inches; brownish-black above with red forewing band and hindwing margin

Butterflies and Moths (Lepidoptera)

Pale Swallowtail
(Papilio eurymedon)

Range: The western U.S.

Wingspan 2.5–3.5 inches; wings whitish with bold black stripes and borders; hindwing with a single long tail

Black Swallowtail
(Papilio polyxenes)

Range: The eastern U.S. and the desert Southwest

Wingspan 2.5–4.2 inches; wings black; male with prominent yellow spot bands; female with reduced yellow and increased blue scaling on hindwing; short hindwing tail; abdomen black with yellow spots

Pipevine Swallowtail
(Battus philenor)

Range: Southern two-thirds of the U.S.

Wingspan 2.75–4.0 inches; overall black; male with iridescent blue scaling on upper side of the hindwing; female duller black with pale marginal spots; hindwing below black with iridescent blue scaling and prominent bright orange spots

Western Tiger Swallowtail
(Papilio rutulus)

Wingspan 2.75–4.0 inches; wings yellow with bold black stripes and borders; hindwing with a single long tail

Range: The western U.S.

Cloudless Sulphur
(Phoebis sennae)

Range: The southern half of the U.S.

Wingspan up to 3.0 inches; wings greenish-yellow with varying amounts of brownish markings

Gulf Fritillary
(Agraulis vanillae)

Range: The southern U.S.

Wingspan up to 3.0 inches; orange above with black markings; brownish-orange below with silvery spots

Anise Swallowtail
(Papilio zelicaon)

Range: The western U.S.

Wingspan up to 3.2 inches; wings black with broad yellow band and a single hindwing tail

Monarch
(Danaus plexippus)

Range: Throughout the U.S. and southern Canada

Wingspan up to 4.0 inches; orange with black veins and wing borders and numerous white spots

Great Spangled Fritillary
(Speyeria cybele)

Wingspan up to 4.0 inches; bright orange with black markings; orange-brown below with metallic silver spots

Range: The northern two-thirds of U.S.

Mourning Cloak
(Nymphalis antiopa)

Range: Throughout the U.S.

Wingspan up to 4.0 inches; velvety black above with a broad irregular yellow border and blue spots; dull black below

Butterflies and Moths (Lepidoptera)

Butterflies and Moths (Lepidoptera)

True Flies (Diptera)

Eastern Tiger Swallowtail
(Papilio glaucus)

Wingspan up to 5.5 inches; wings yellow with black stripes and borders and single hindwing tail; some females are primarily black

Range: The eastern U.S.

Two-tailed Swallowtail
(Papilio multicaudata)

Wingspan up to 6.0 inches; wings yellow with black stripes and borders; hindwing with two prominent tails

Range: The western U.S.

Pearl Crescent
(Phyciodes tharos)

Range: East of the Rocky Mountains

Wingspan 1.25–1.6 inches; wings above orange with dark bands, spots, and borders; hindwing below tan with dark patch at border with a pale crescent spot

Milbert's Tortoiseshell
(Aglais milberti)

Wingspan 1.8–2.5 inches; forewing squared off; wings above dark with broad yellow-orange band; wings with irregular margins; wings below with a bark-like pattern

Range: The western U.S., the Northeast,, and parts of the Midwest

Flesh Fly
(Family Sarcophagidae)

0.4–0.6 of an inch long; gray body with longitudinal black stripes; bristled abdomen and reddish eyes

Range: Throughout the U.S.

Bee Fly
(Family Bombyliidae)

0.4–0.75 of an inch long; hairy body with wings held outward at rest; long forward-pointing proboscis; resembles a small bee

Range: Throughout the U.S.

Long-horned Beetle
(Family Cerambycidae)

0.5–2.5 inches long; variable in color; slender oval body with long, noticeable antennae

Range: Throughout the U.S.

Jagged Ambush Bug
(Genus *Phymata*)

0.4–0.5 of an inch long; variable mottled color pattern; violin-shaped body with grasping front legs

Range: Throughout the U.S.

Metallic Green Sweat Bee
(Genus *Agapostemon*)

Up to 0.5 of an inch long; head and thorax metallic green; wings dark; abdomen metallic green or yellow with black bands

Range: Throughout the U.S.

Yellow Jacket
(Genus *Vespula*)

Up to 0.6 of an inch long; body with yellow and black markings; narrow, amber wings

Range: Throughout the U.S.

True Flies (Diptera)

Beetles (Coleoptera)

True Bugs (Hemiptera)

Wasps, Bees, Ants and Sawflies (Hymenoptera)

European Honey Bee
(Apis mellifera)

Up to 0.75 of an inch long; fuzzy appearance; large black eyes, two pairs of wings; black and golden-orange striped abdomen

Range: Throughout the U.S.

Bumble Bee
(Genus Bombus)

Range: : Throughout the U.S.

Up to 1.5 inches long; fuzzy, robust body; prominent black-and-yellow pattern

Flower Crab Spider
(Genus Misumena)

Range: Throughout the U.S.

Up to 0.5 of an inch long; highly variable in color; flattened body with rounded abdomen; eight legs, with the two front pairs noticeably longer than hind legs; females much larger than males

Green Lynx Spider
(Peucetia viridans)

Up to 0.85 of an inch long; bright green body; eight legs bearing black spines and spots

Range: Throughout the southern U.S.

Common Wood-Nymph
(Cercyonis pegala)

Wingspan up to 2.8 inches; brown with two large yellow-rimmed eyespots; some with a yellow patch surrounding eyespots

Range: Throughout much of the U.S.

Question Mark
(Polygonia interrogationis)

Range: East of the Rocky Mountains

Wingspan up to 3.0 inches; orange above with black spots; forewing squared off; hindwing with stubby tail; variable brown below with a dead leaf pattern

Viceroy
(Limenitis archippus)

Wingspan up to 3.2 inches; orange with black veins and borders; hindwing with a central black line

Range: Throughout most of the U.S., except the West Coast

Leaf Beetle
(Family Chrysomelidae)

0.04–0.7 of an inch long; highly variable in color and pattern; typically shiny

Range: Throughout the U.S.

Weevil
(Family Curculionidae)

0.10–0.30 of an inch long; highly variable; typically dark, oval to pear-shaped body; distinctive snout extends far off the head

Range: Throughout the U.S.

Butterflies and Moths (Lepidoptera)

Beetles (Coleoptera)

ON VEGETATION

Beetles (Coleoptera)

Lady Beetle
(Family Coccinellidae)

Range: Throughout the U.S.

0.25–0.40 of an inch long; oval shape; variable pattern; orange to red with black spots

Japanese Beetle
(*Popillia japonica*)

0.3 to 0.5 of an inch long; metallic green with copper-colored wing coverings and white tufts along the abdomen; often feeds in groups

Range: Mostly in the eastern U.S.; found sporadically westward

Hover Fly
(Family Syrphidae)

0.25-0.75 inches long; highly variable, often resembling bees with yellow and black pattern; large eyes and generally unmarked transparent wings

Range: Throughout the U.S.

Non-biting Midge
(Family Chironomidae)

Range: Throughout the U.S.

0.04–0.4 of an inch long; narrow body with long legs and fern-like antennae; resembles a mosquito

Long-Legged Fly
(Family Dolichopodidae)

0.05–0.35 of an inch long; small, often metallic body; bright eyes; mostly transparent wings and long, thin legs

Range: Throughout the U.S.

True Flies (Diptera)

Range: Throughout the U.S.

Robber Fly
(Family Asilidae)

0.3–1.0 inch long; highly variable in appearance; typically slender body with a tapered abdomen; large eyes; two wings; long, bristly legs

Range: Throughout the U.S.

Green Bottle Fly
(*Lucilia sericata*)

0.4–0.55 of an inch long; stout metallic green body; large eyes and two transparent wings

True Flies (Diptera)

Aphid
(Family Aphididae)

0.10–0.15 of an inch long; oval body with pair of pipe-like projections on the abdomen; variable color from green and yellow to orange and even black

Range: Throughout the U.S.

Planthopper
(Superfamily Fulgoroidea)

Range: Throughout the U.S.

0.10–0.25 of an inch long; variable in color; white, brown, or green; broad wings held vertically; some distinctively wedge-shaped

Leafhopper
(Family Cicadellidae)

0.15–0.75 of an inch long; narrow wedge-shaped body; highly variable color, including green, brown, or brightly patterned

Range: Throughout the U.S.

True Bugs (Hemiptera)

Assassin Bug
(Family Reduviidae)

0.2–1.5 inches long; highly variable in color; oblong body with long legs and elongated beak for stabbing prey

Range: Throughout the U.S.

Treehopper
(Family Membracidae)

0.25–0.50 of an inch long; highly variable, from green to brown to multicolored; typically resembles small leaves or thorns

Range: Throughout the U.S.

Large Milkweed Bug
(Oncopeltus fasciatus)

0.4–0.6 of an inch long; elongated orange-and-black body; found on milkweed plants

Range: Throughout the U.S.

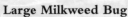

Green Stink Bug
(Family Pentatomidae)

0.5–0.7 of an inch long; bright green body that is flattened and shield-shaped

Range: Throughout the U.S.

Cicada
(Family Cicadidae)

1.0–2.0 inches long; stout, robust body; widely separated eyes; clear transparent wings; resembles a large fly

Range: Throughout the U.S.

True Bugs (Hemiptera)

ON VEGETATION

Jumping Spider
(Family Salticidae)

0.15–0.6 of an inch long; highly variable in color; stout, hairy body with two of its eight eyes looking straight ahead; eight legs

Range: Throughout the U.S.

Black and Yellow Garden Spider
(*Argiope aurantia*)

0.75–2.5 inches long; abdomen egg-shaped with black and yellow markings; eight black legs marked with yellow or red; females much larger than males

Range: Throughout the U.S.

Praying Mantis
(Family Mantidae and Liturgusidae)

2.0–4.0 inches long; long, slender green to brown body; triangular head; enlarged front legs for grasping prey; immatures are wingless; adults have membranous wings

Range: Throughout the U.S.

Walking Stick
(Order Phasmida)

Range: Throughout the U.S.

3.0–3.6 inches long; long, slender brown or green body; long legs and antennae; resembles a small twig

Ichneumonid Wasp
(Family Ichneumonidae)

0.1–1.5 inches long; highly variable in color; long, slender body with long antennae and legs

Range: Throughout the U.S.

Wood Tick
(Family Ixodidae)

0.12–0.18 of an inch long; flattened dark brown to reddish oblong body, often with some lighter markings; eight legs and forward-pointing mouthparts

Range: The eastern two-thirds of the U.S. and on the West Coast

Spiders (Araneae)

Mantises (Mantodea)

Walking Sticks (Phasmida)

Wasps, Bees, Ants, and Sawflies (Hymenoptera)

Ticks (Ixodida)

Rosy Maple Moth
(*Dryocampa rubicunda*)

Range: The eastern U.S.

Wingspan up to 2.0 inches; bright pink and yellow wings; fuzzy yellow body

Io moth
(*Automeris io*)

Range: From the Great Plains east

Wingspan up to 3.0 inches; yellow male, brown female; hindwing with a large eyespot

Underwing Moth
(Genus *Catocala*)

Range: Throughout the U.S.

Wingspan up to 3.0 inches; variable; typically dark forewings with bark-like pattern; hindwings with colorful bands

White-lined Sphinx
(*Hyles lineata*)

Wingspan up to 3.5 inches; stout elongated body; elongated brown forewings with pale central stripe; hindwing pink with dark borders

Range: Throughout the U.S.

Giant Leopard Moth
(*Hypercompe scribonia*)

Wingspan up to 3.5 inches; elongated white wings with mix of solid black and hollow black spots

Range: The eastern U.S.

Royal Walnut Moth
(*Citheronia regalis*)

Wingspan 3.9–6.25 inches; wings grayish with orange veins and cream spots; body orange with a creamy yellow pattern

Range: Primarily the southeastern U.S. into the Northeast and the Midwest

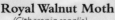

Butterflies and Moths (Lepidoptera)

Cecropia Silkmoth
(Hyalophora cecropia)

Range: The eastern U.S.

Wingspan 4.25–5.75 inches; wings brown with pale margins, a red and white band, and a prominent central crescent light spot; body patterned red and white

Luna Moth
(Actias luna)

Range: The eastern U.S.

Wingspan up to 4.5 inches; light green with long hindwing tails and a furry white body

Polyphemus Moth
(Antheraea polyphemus)

Wingspan up to 5.8 inches; tan to reddish-brown with a large hindwing eyespot

Range: Throughout the U.S.

Imperial Moth
(Eacles imperialis)

Range: The eastern U.S.

Wingspan up to 6.8 inches; elongated yellow wings with varying degree of purplish-brown markings

Giant Water Bug
(Family Belostomatidae)

Up to 2.25 inches long; dull brown; wings overlap to form a distinctive X-like pattern on the back; forward-facing, pincer-like front legs

Range: Throughout the U.S.

Eastern Dobsonfly
(Corydalus cornutus)

4.0–5.5 inches long; brown body with membrane-like gray-brown wings; males have large curved jaws

Range: From the Rocky Mountains and east

Butterflies and Moths (Lepidoptera)

True Bugs (Hemiptera)

Alderflies, Dobsonflies and Fishflies (Megaloptera)

Beetles (Coleoptera)

June Bug/May Beetle
(Genus *Phyllophaga*)

Range: Throughout the U.S.

Up to 1.0 inch long; stout, unmarked, shiny tan-to-reddish-brown oblong body

Click Beetle
(Family Elateridae)

Up to 1.2 inches long; typically brown to black; elongated often dull and unmarked body

Range: Throughout the U.S.

Ten-lined June Beetle
(*Polyphylla decemlineata*)

Range: Western U.S.

Up to 1.25 inches long; brown oval body; vertical white stripes and noticeable clubbed antennae

True Flies (Diptera)

Crane Fly
(Family Tipulidae)

Up to 2.5 inches long; slender body with extremely long, thin legs; two elongated wings

Range: Throughout the U.S.

Antlions, Lacewings, and Mantidflies (Neuroptera)

Green Lacewing
(Family Chrysopidae)

Range: Throughout the U.S.

Up to 0.7 of an inch long; light green slender body; long antennae and four transparent wings with green veins

Ant Lion
(Family Myrmeleontidae)

Up to 1.5 inches long; long, slender body with short antennae; large transparent wings

Range: Throughout the U.S.

Mayfly
(Order Ephemeroptera)

Range: Throughout the U.S.

Up to 1.0 inch long; narrow pale to dark body; long legs; four wings held upright over back; two to three long filaments off abdomen

Caddisfly
(Order Trichoptera)

Up to 1.25 inches long; narrow body with transparent wings; long thin antennae

Range: Throughout the U.S.

Stonefly
(Order Plecoptera)

Range: Throughout the U.S.

Up to 1.6 inches long; dull brown; elongated and flattened body with transparent wings; long legs and two long tail filaments

Mayflies (Ephemeroptera)

Caddisflies (Trichoptera)

Stoneflies (Plecoptera)

IN OR NEAR WATER

True Bugs (Hemiptera)

Water Strider
(Family Gerridae)

0.4–0.5 of an inch long; brown narrow body with long middle and hind legs; found on the water's surface

Range: Throughout the U.S.

Range: The eastern U.S.

Eastern Amberwing
(Perithemis tenera)

0.8–1.0 inch long; brown body with transparent orange-to-amber wings

Pond Damselfly
(Family Coenagrionidae)

1.0–1.70 inches long; variable color; thin elongated body with transparent wings together over the back; head with large eyes

Range: Throughout the U.S.

Common Whitetail
(Plathemis lydia)

1.6–1.9 inches long; stout body; clear wings with wide black central band; chalky white abdomen

Range: Throughout the U.S.

Twelve-Spotted Skimmer
(Libellula pulchella)

2.0–2.25 inches long; wings transparent with alternating dark and white spots; grayish abdomen

Range: Throughout the U.S.

Dragonflies and Damselflies (Odonata)

Common Green Darner
(*Anax junius*)

2.75–3.2 inches long; thorax green
with long, slender blue abdomen;
four transparent wings

Range: Throughout the U.S.

Dragonflies and
Damselflies (Odonata)

Range: Throughout the U.S.

Whirligig Beetle
(Family Gyrinidae)

0.25–0.50 of an inch long; shiny black,
elliptical body; swims erratically on
surface of water; often seen in groups

Beetles (Coleoptera)

True Flies (Diptera)

Mosquito
(Family Culicidae)

Typically less than 0.25 of an
inch long; narrow, often striped
body with two wings; long thin
legs and a prominent proboscis

Range: Throughout the U.S.

Beetles (Coleoptera)

Firefly
(Family Lampyridae)

Range: Eastern U.S.

0.5–0.8 of an inch long; flattened dark
body with orange and yellow markings;
underside tip of abdomen is pale and
produces light

Paper Wasp
(Family Vespidae)

Range: Throughout the U.S.

Up to 1.0 inch long; variable; slender brown to black body with yellow markings; narrow waist; pointed abdomen and elongated wings

Wasps, Bees, Ants, and Sawflies (Hymenoptera)

Sowbug/Pillbug
(Family Oniscidae and Armadillidiidae)

Range: Throughout the U.S.

0.25–0.5 of an inch long; dark gray to brown; oval body with plate-like segments and seven pairs of small legs

Woodlice (Isopoda)

Ground Beetle
(Family Carabidae)

Range: Throughout the U.S.

0.10–1.25 inches long; highly variable; typically black, brown, or metallic in color; dull or shiny; long legs, prominent jaws

Tiger Beetle
(Family Cicindelidae)

0.5–0.9 of an inch long; shiny, often iridescent, pattered body; large eyes; long legs; prominent jaws

Range: Throughout the U.S.

Metallic Wood-Boring Beetle
(Family Buprestidae)

0.5–1.5 inches long; streamlined, bullet-shaped body; flat head, short antennae; tapered rear; shiny; often metallic colors

Range: Throughout the U.S.

Spiders (Araneae)

Wolf Spider
(Family Lycosidae)

Range: Throughout the U.S.

0.75–1.3 inches long; hairy gray to dark brown body with darker markings; eight long legs; resembles a small tarantula

Millipedes (Julida)

Millipede
(Order Julida)

0.5–6.0 inches long; dark, smooth, cylindrical, and segmented worm-like body with many small legs

Range: Throughout the U.S.

Centipede
(Order Chilopoda)

1.0–1.5 inches long; typically tan to dark brown; segmented worm-like body with numerous long legs

Range: Throughout the U.S.

Termites (Blattodea)

American Cockroach
(*Periplaneta americana*)

Range: Primarily the southern U.S.

1.0–2.8 inches long; oblong reddish-brown body; spiny legs and long thin antennae

Vinegaroon
(Family Thelyphonidae)

1.5–2.5 inches long; dark brown to black; six walking legs; pair of long modified front legs resemble antennae; two enlarged front pincers; long, thin tail filament

Range: Primarily Southwest and Florida

Scorpion
(Order Scorpiones)

0.9–3.0 inches long; typically brown to black; eight legs; prominent grasping claw-like pedipalps toward the front of its body; long segmented and curved tail with stinger on the end

Range: Primarily the Southwest and Florida

Snail
(Order Pulmonata)

Highly variable size from 0.2–8.0 inches long; brown to dull-colored fleshy body with elongated retractable tentacles off the head; hard, often patterned, shell on its back

Range: Throughout the U.S.

Slug
(Order Soleolifera)

Range: Throughout the U.S.

1.0–4.0 inches long; brown to dull-colored; fleshy body is often patterned; elongated stalk-like retractable tentacles protruding from head; lacks a hard shell

Earwig
(Order Dermaptera)

Range: Throughout the U.S.

0.5–0.75 of an inch long; elongated reddish-brown body; prominent pincers off the end of the abdomen

Daddy Longlegs
(Order Opiliones)

Round brownish body 0.15–0.3 of an inch long with much longer thin legs

Range: Throughout the U.S.

ON GROUND/STRUCTURES

Tree Cricket
(Family Gryllidae)

0.4–0.8 of an inch long; green slender body; long hind legs and antennae; two pairs of membranous wings

Range: Throughout the U.S.

Katydid
(Family Tettigoniidae)

0.5–2.5 inches long; generally green body with large hind legs; long antennae; may have large wings; females have curved ovipositor at end of abdomen

Range: Throughout the U.S.

Field Cricket
(Genus *Gryllus*)

0.75–1.2 inches long; brown to black with membranous wings; enlarged hind legs; long antennae; two prominent tail filaments

Range: Throughout the U.S.

Carolina Locust
(*Dissosteira carolina*)

1.2–2.0 inches long; mottled gray to brown elongated body; adults with membranous wings

Range: Throughout the U.S.

Grasshopper
(Family Acrididae)

0.4–4.0 inches long; highly variable color and pattern; elongated body with long hind legs modified for jumping; adults with membranous wings

Range: Throughout the U.S.

Ant
(Family Formicidae)

Range: Throughout the U.S.

0.04–0.75 of an inch long; variable in color; typically yellow, reddish, brown to black; elongated body with distinct head, thorax, and abdomen; antennae elbowed

Range: Throughout the U.S.

Cellar Spider
(Family Pholcidae)

Body 0.25–0.35 of an inch long; small oblong gray to brown body; eight very long, thin legs

bees, wasps, and ants (Hymenoptera)

Tussock Moth Caterpillar
(Genus *Orgyia*)

1.0–1.5 inches long; hairy, gray to cream body with two long black hair tufts extending forward past the head and one off the rear; four compact brushy hair tufts on the back

Range: The eastern U.S.

Bagworm
(Family Psychidae)

1.0–2.0 inches long; brown, cocoon-like structure with bits of leaves and twigs that conceals the caterpillar inside; hangs from branches

Range: Throughout the U.S.

Monarch Caterpillar
(*Danaus plexippus*)

1.25–2.0 inches long; body striped with alternating bands of black, white, and yellow; a pair of long, black filaments on both ends

Range: Throughout the U.S.

Black Swallowtail Caterpillar
(*Papilio polyxenes*)

1.75–2.2 inches long; green with black bands and yellow-orange spots

Range: The eastern U.S.

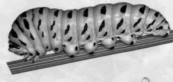

Banded Woolly Bear Caterpillar
(*Pyrrharctia isabella*)

1.8–2.25 inches long; densely fuzzy; banded with black at both ends and reddish-brown in the middle

Range: Throughout the U.S.

Polyphemus Moth Caterpillar
(Antheraea polyphemus)

2.3–2.9 inches long; bright green body with thin yellow vertical stripes and a brown head

Range: Throughout the U.S.

Tomato Hornworm
(Manduca quinquemaculata)

3.0–4.0 inches long; bright green body with seven white stripes on the side; a prominent curved horn off the back

Range: Throughout most of the U.S.

Cecropia Moth Caterpillar
(Hyalophora cecropia)

4.0–4.5 inches long; bluish-green body with bright blue, red, and yellow tubercles (outgrowths), each with short black spines

Range: The eastern U.S.

Hickory Horned Devil
(Citheronia regalis)

4.5–5.5 inches long; greenish-blue body with an orange head; long, curved red-and-black horns

Range: Primarily the southeastern U.S. into the Northeast and the Midwest

JARET C. DANIELS

Longtime nature photographer and entomologist at the University of Florida, Jaret Daniels specializes in insect ecology and conservation. He has authored numerous scientific papers, popular articles, and books on gardening, wildlife conservation, insects, and butterflies, including butterfly field guides for Florida, the Carolinas, Ohio, and Michigan. His most recent titles include *Vibrant Butterflies: Our Favorite Visitors to Flowers and Gardens* and *Backyard Bugs: An Identification Guide to Common Insects, Spiders, and more*. He lives in Gainesville, Florida, with his wife, Stephanie.

Adventure Quick Guides

Quick & Easy Bug ID

Organized by where the bugs are generally found, such as at lights or on flowers

Simple and convenient—narrow your choices by group, then size, and view just a few species at a time

- Pocket-size format—easier than laminated foldouts

- Professional photos showing key traits

- Ants, bees, beetles, butterflies, dragonflies, spiders, wasps, and much more

- Fascinating introduction to insect anatomy

- Expert author is a noted entomologist

Improve your identification skills with this handy field guide

$9.95

Adventure PUBLICATIONS

Adventure Publications
330 Garfield Street South
Cambridge, Minnesota 55008
(800) 678-7006
www.adventurepublications.net

NATURE/INSECTS

ISBN 978-1-59193-818-7

5 0 9 9 5

9 781591 938187